**Sharp Series**

# Color Your Way to Math Volume 3

**Conceptual Learning
A Beginner's Visual Guide
to Pre-K to Grade 2 Math**

**Lex Sharp**

**Bridget Koteles**

Fields of Code Inc.
Calgary, Alberta
www.fieldsofcode.ca

Copyright © 2019 Fields of Code Inc. All Rights Reserved.

Published by Lex Sharp

Art Editor: Bridget Koteles

Fields of Code Inc.
Calgary, Alberta
Canada
www.fieldsofcode.ca

No part of this publication may be reproduced in any form or by any means, including scanning, photocopying, or otherwise without prior permission of the copyright holder.

ISBN: 978-1-689-213905

FIRST EDITION

# Table of Contents

In This Booklet ..................................................................................... 1

Worksheet 1 ........................................................................................ 2

Worksheet 2 ........................................................................................ 4

    Three in Four, What's That? ............................................................ 4

    Draw Your Own Shapes .................................................................. 6

Worksheet 3 ........................................................................................ 8

Worksheet 4 ...................................................................................... 10

Worksheet 5 ...................................................................................... 12

Worksheet 6 ...................................................................................... 14

Worksheet 7 ...................................................................................... 16

    Conclusion: The Number of Cars Was Even! ................................. 24

Worksheet 8 ...................................................................................... 26

    Conclusion: The Number of Toys Was Odd! .................................. 26

Worksheet 9 ...................................................................................... 28

Worksheet 10 .................................................................................... 30

Worksheet 11 .................................................................................... 32

Worksheet 12 .................................................................................... 34

Worksheet 13 .................................................................................... 36

Worksheet 14 .................................................................................... 38

Worksheet 15 .................................................................................... 40

Worksheet 16 .................................................................................... 48

Answers ............................................................................................. 51

Errata and Feedback ......................................................................... 59

Other Books in This Series ................................................................ 59

# In This Booklet...

This volume is part of a mathematical series for preschoolers to grade 2.

Each booklet takes a bite-sized approach to avoid overwhelming the student. Volumes were design to have a size and font that is comfortable to preschool aged children.

This volume covers the following topics:
- linear and circular patterns,
- preparing for fractions: using 1 in 4, 2 in 4, and so on,
- writing and recognizing numbers,
- using ordinal numbers,
- using a ruler,
- even and odd numbers,
- paying attention to a series of instructions,
- the logical operators OR,
- addition,
- understanding two equal sides, finding how many are missing from one side: a precursor to equations,
- recognizing the net of a cube.

Lex Sharp

# Worksheet 1

The cars on the opposite page are forming patterns.

Patterns use sets of items that appear repeatedly.

A question mark was added for a missing car in every row.

Color the box that shows the missing car below each pattern.

Find a group of cars that repeats in each row. Circle it and color the cars in it.

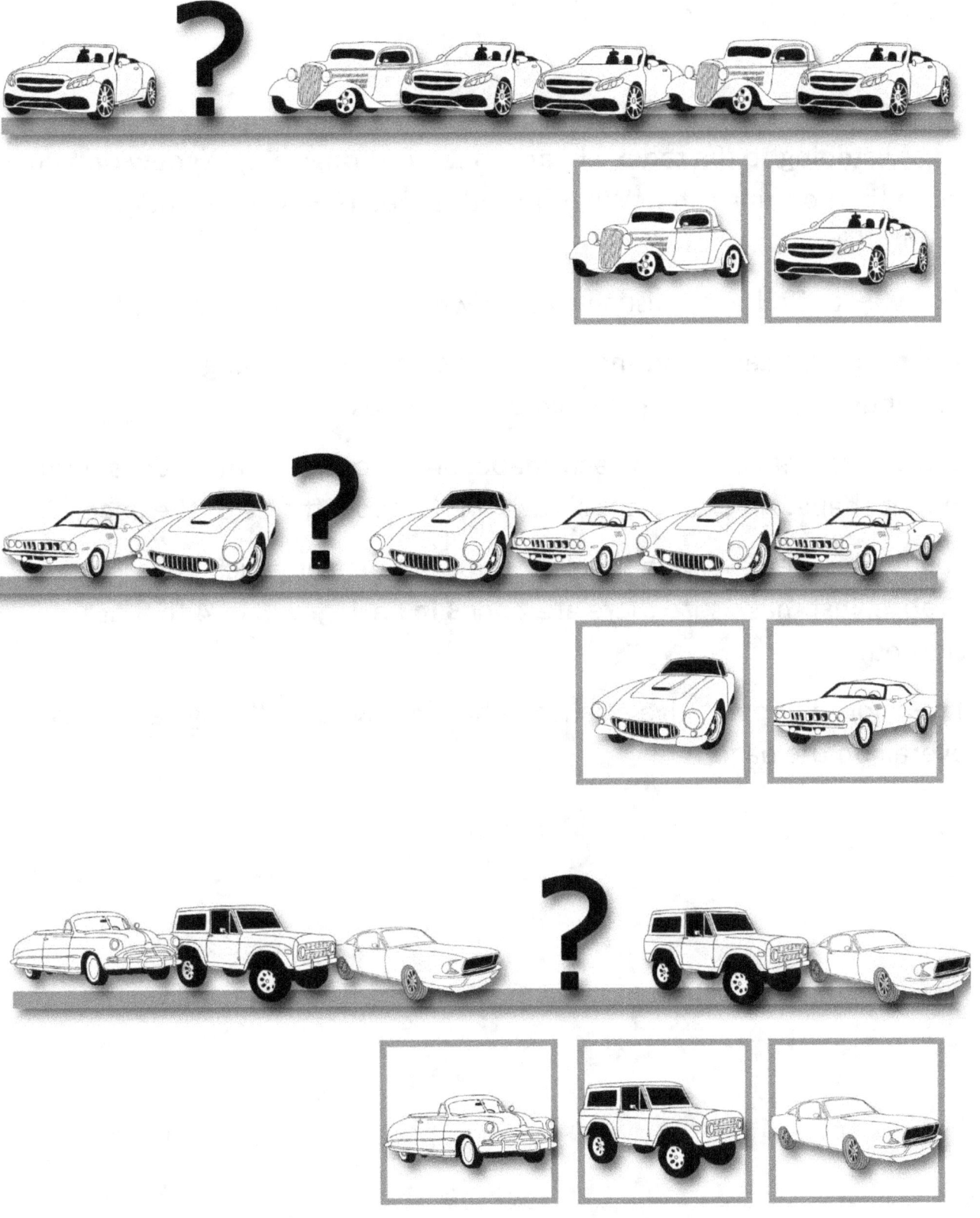

# Worksheet 2

## Three in Four, What's That?

I was listening to the radio, I heard that 3 in 4 dogs like to chew on bones more than playing catch. What does this mean? How many dogs were counted in this story?

One way to find out is visually, as shown on the opposite page.

Any **four dogs** can be arranged in circles, squares, rectangles, rhombuses, ovals, etc., as shown on the opposite page.

Use a different color with each shape. Shade only 3 of the 4 dogs found in that shape. An example is shown on the top left side of the page.

Follow this idea everywhere else. It doesn't matter which dogs you choose, just make sure there are **only 3 in each group** of 4 that are colored.

The colored puppies give a sense of how many "**3 in 4**" dogs are present overall on the page.

Color Your Way to Math, Volume 3

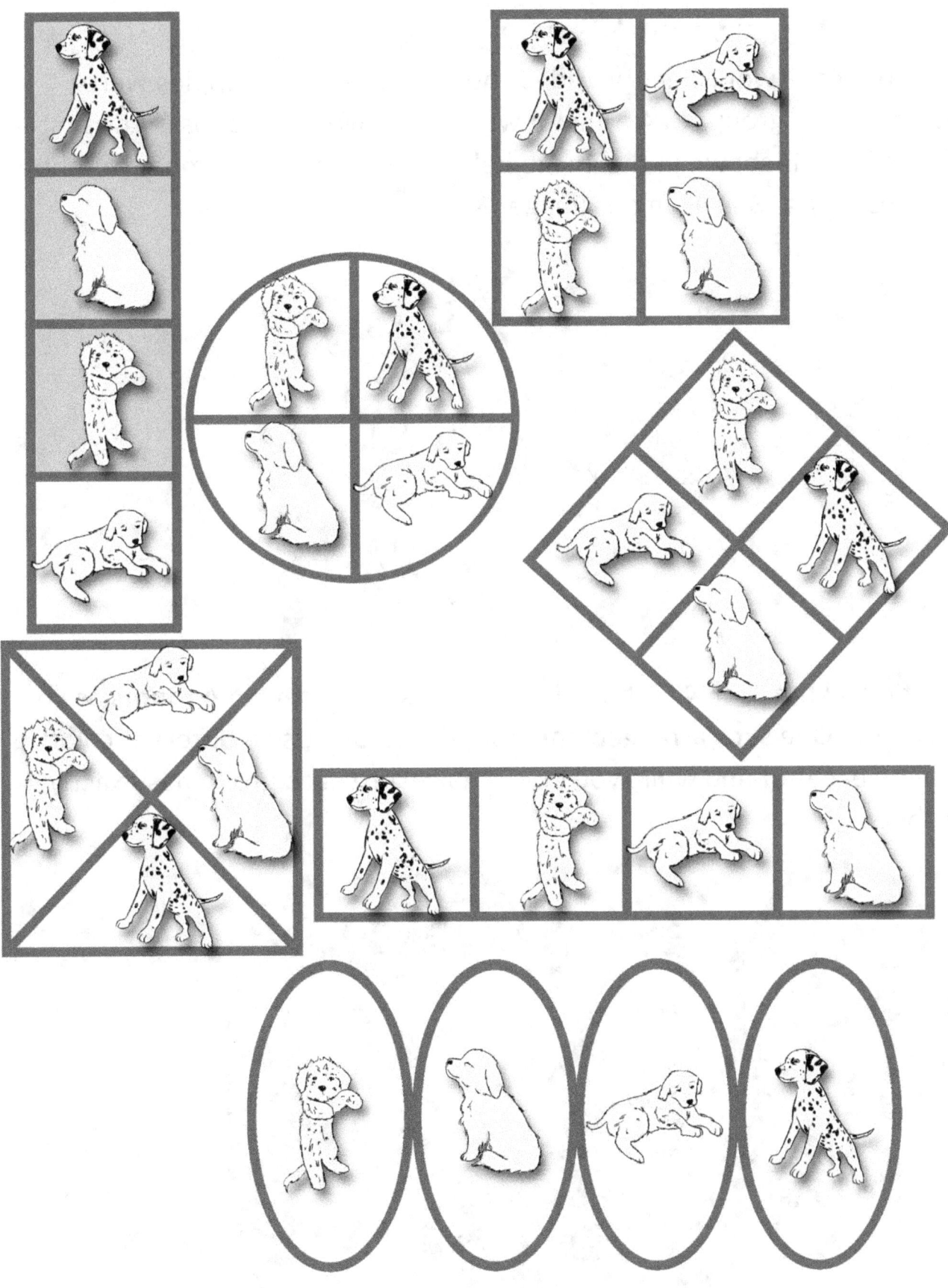

# Draw Your Own Shapes

The previous worksheet provided the shapes, and all puppies were organized in groups of **4**. What if we didn't have the groups in the first place? This is shown in the image below. Let's find the number of robots representing "**3 in 4**" in this image.

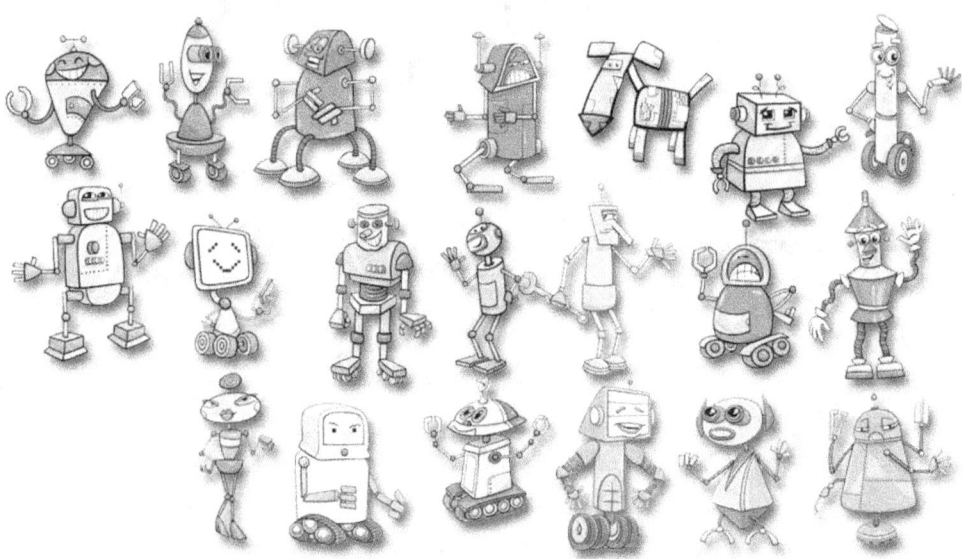

To find all the "**3 in 4**" robots, first circle as many sets of **4** robots as you can find. One such arrangement is shown below. Use a different color to shade each set and while you do so make sure each group has exactly 4 robots.

Now that all robots are organized in sets of **4** it's easy to mark **3** of them in each group. Let's cross out 3 robots in each set. The **X** signs below depict how many "**3 in 4**" were found overall.

Every group has 1 robot left over that is not crossed out. These are circled below. Let's call these "**1 in 4**" robots, because there is **1** such in every group of **4**.

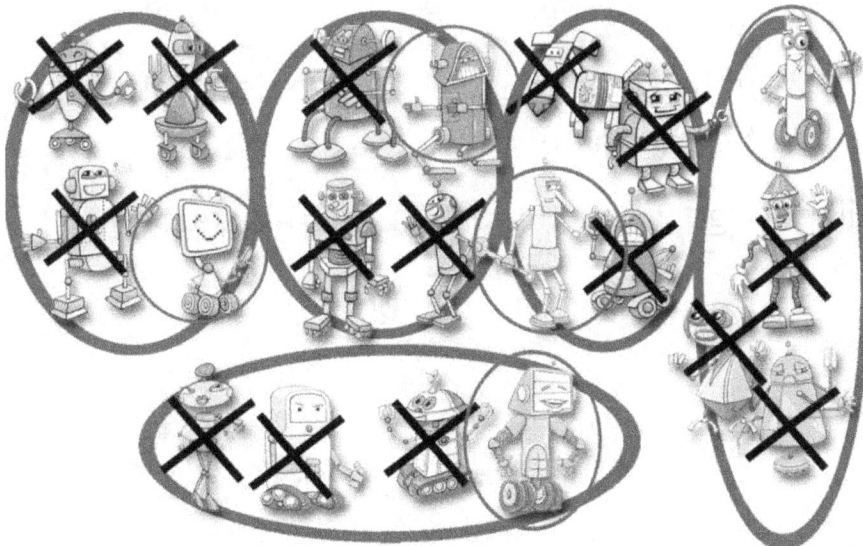

Notice there are a lot more robots crossed out than circled. There are more robots of the "**3 in 4**" kind than there are those of the "**1 in 4**" kind.

# Worksheet 3

You must be able to count to 20 to solve this worksheet.

You will need a **ruler** to draw straight lines between the dots.

Every dot in the picture has one or more numbers attached to it. For example, at the top, a dot is both a **1** and a **2**, as shown by the two arrows below.

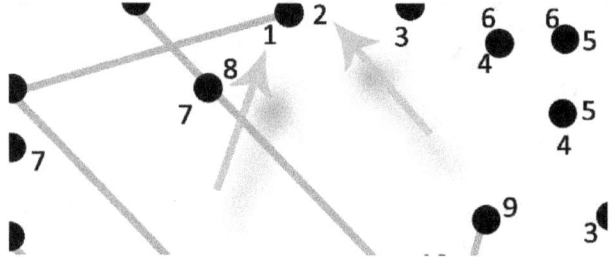

Every number between 1 and 20 appears **exactly twice** in the image.

To play this game, follow these steps:

- Find **two dots** that have **the same numbers**, they make a pair.
- Place a ruler between the two dots you've found and draw a straight line between them.
- Repeat these steps until you can no longer find two dots with the same numbers. Can you find all the pairs?

When done, a pattern will emerge. Color the pattern with beautiful colors.

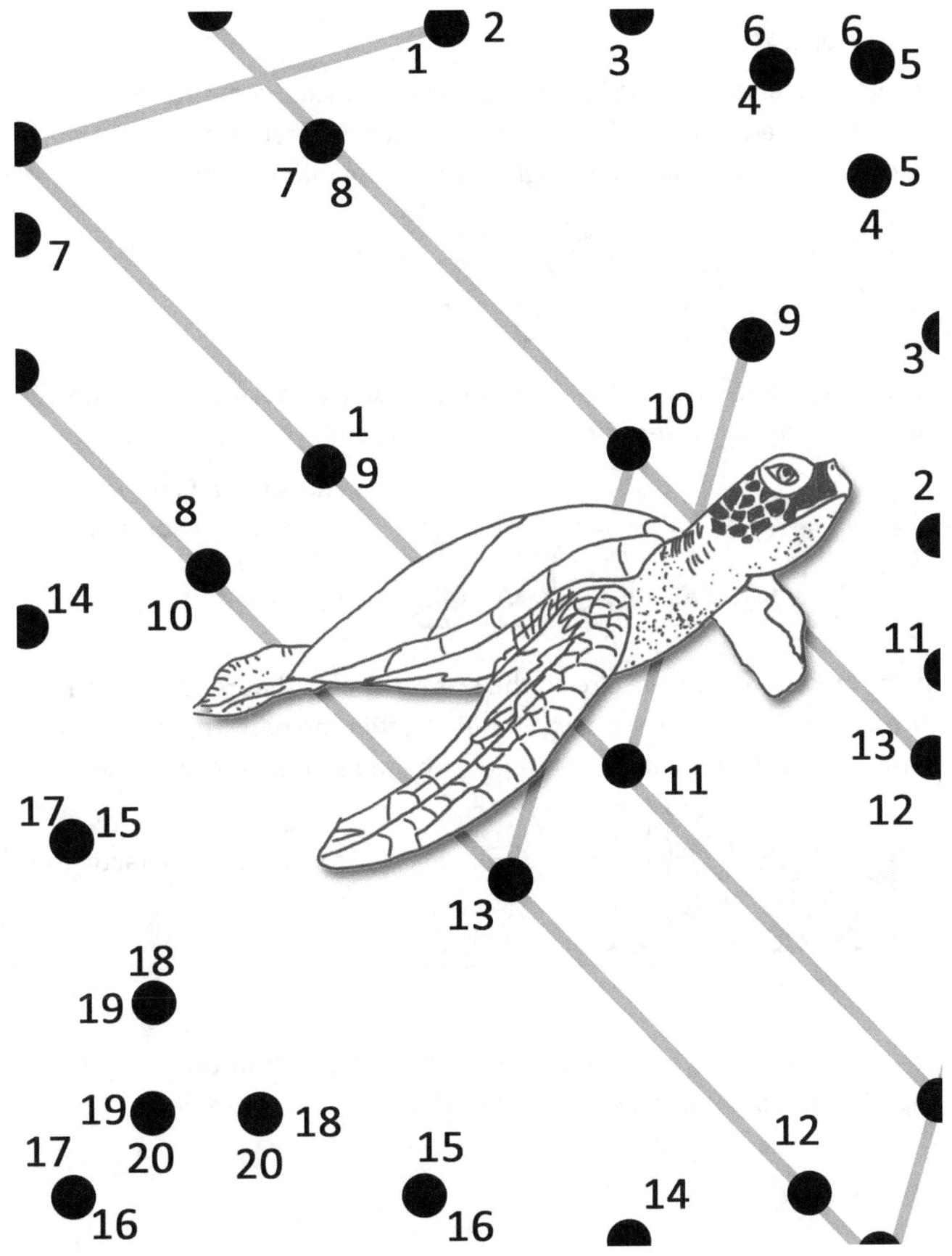

# Worksheet 4

The shapes, lines, and the letter S below were mixed together to build the shape of a cat. From this basic shape, two new types of cats were made, as shown below on the right, by coloring different parts.

When these two cats are repeated, they make a pattern as shown below. The repeating parts were circled with a wavy line. I didn't have to

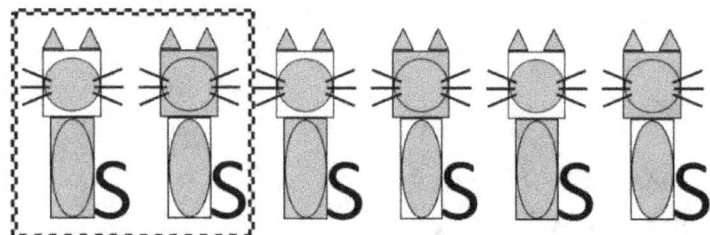

choose the first two. I could have chosen any other two that appear in this order.

In another pattern I have circled the last three cats that repeat. I made sure the first cat in my circle was the first cat in the pattern, the second in my circle was the second in the pattern, and so on. The important

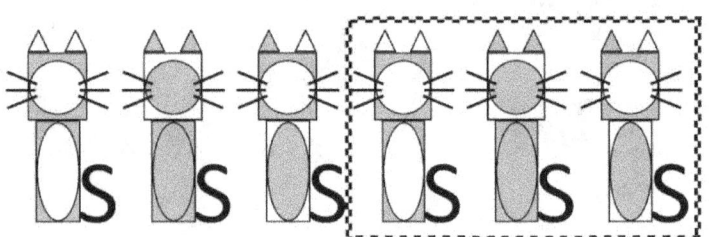

thing is that the circled cats repeat in exactly the same order.

Use colors to make your own unique pattern in each row on the opposite page. Circle a set that repeats to make the pattern in every row.

*(The answer key gives a few of many possible solutions. There are infinite correct answers and I bet yours is great!).*

Color Your Way to Math, Volume 3

# Worksheet 5

Make your own pattern.

A pattern doesn't have to be lined up in rows, it can be circular, or spiral, and so on.

The example on the opposite page (at the top) has several circular patterns in it. Study them to see which parts repeat. A few of these circular patterns were extracted below and straightened to demonstrate.

Can you see where each pattern fits into the larger shape?

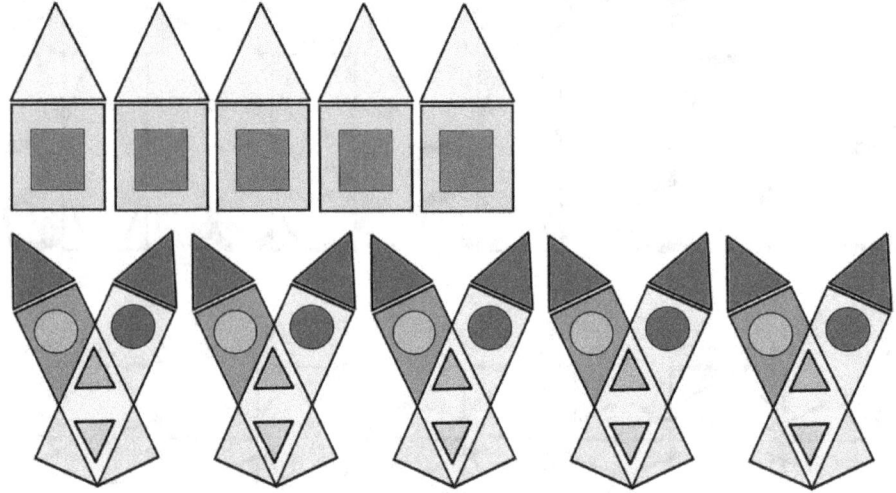

The overall pattern can be opened like this:

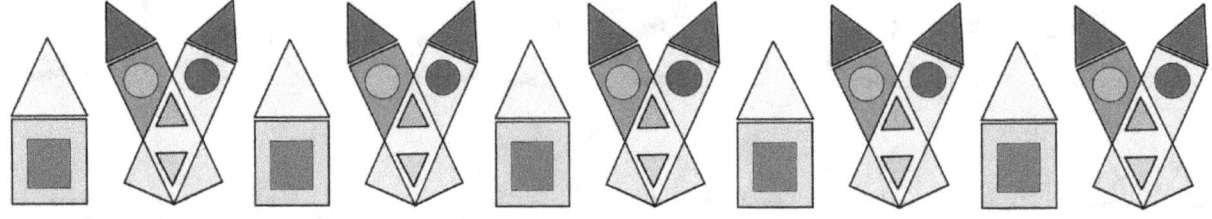

Now it's your turn, color the shape at the bottom of the opposite page in a way that forms a pattern. Make sure your colored pieces repeat.

Color Your Way to Math, Volume 3

13

# Worksheet 6

The circular pattern shown on the opposite page is unfolded at the bottom into a single row. The row is there to help you plan your own circular pattern. Three more rows were added below so you can experiment with what coloring patterns work best.

To line up the pattern circularly you must think about how the first and last elements fit together to close the circle.

Visit the Sharp Series website to download a cutout version of this model. You can then color and position the pieces circularly for a hands-on experiment. Navigate to the main page at: https://sharpseries.ca/cyw/cyw2m.html. Select the *"Download"* link at the top or scroll down to the bottom of the page. Find the **Volume 3** category and download the PDF file tagged as *"Link 04"*.

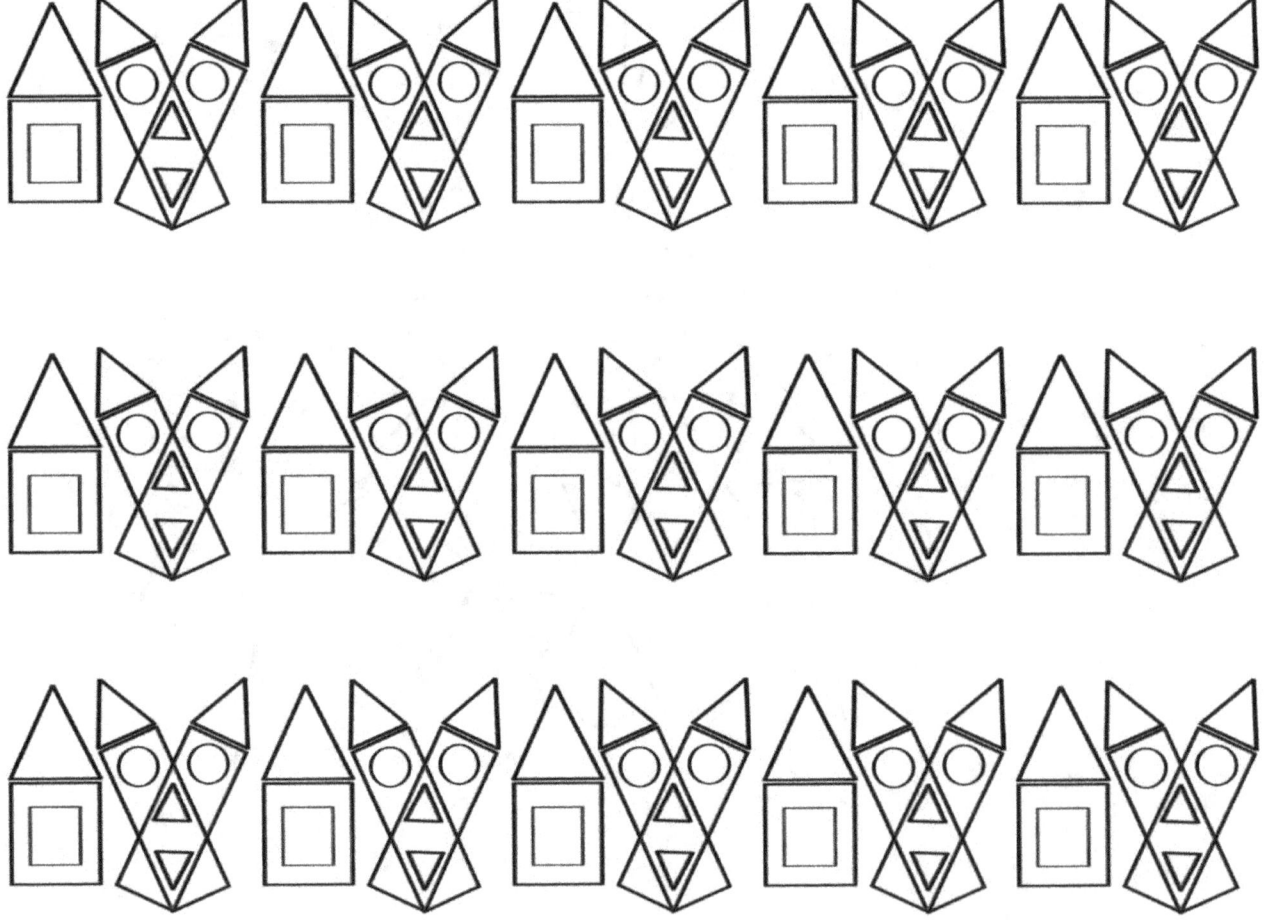

Color Your Way to Math, Volume 3

# Worksheet 7

**Understanding Even and Odd...**

**Part 1**

Britney and Jennifer are about to play with a car collection.

They want to race the cars several times using a unique pair of cars in every race.

They need to split the collection in pairs between them.

For now, color all the cars on the opposite page just for fun while you think about what the best way is to divide the collection.

When finished coloring, move on to **Part 2**.

## Part 2

Let's think only about the first three car races for now.

Two cars that will race together are tagged by the number of the race as shown in the bubbles on the opposite page. This means:

**1** is the tag for the **first** race,

**2** is the tag for the **second**, and

**3** is the tag for the **third** race.

Color the cars assigned to the first three races using one color.

Use a different color to shade in the cars that were **not** assigned to a race yet.

Answer the following questions.

| Question | Answer |
|---|---|
| How many cars has **each** girl picked so far based on the given tags? | |
| How many cars were selected so far **overall**? | |
| What is the total number of cars that were not assigned to a race yet? | |

## Part 3

The girls continue to select cars for races 4, 5 and 6.

Choose a different color for each of these races and shade every pair that goes together. Race numbers are shown in bubbles.

Remember, the cars tagged in Part 2 and the newly colored ones are already selected to race. Answer the following questions.

| Question | Answer |
| --- | --- |
| How many cars has **each** girl picked so far? | |
| How many cars were selected so far **overall**? | |
| What is the total number of cars that were not assigned to a race yet? | |

## Part 4

The girls continue to pick cars for the last race numbered 7.

Choose a new color for this race and shade the last pair that goes together.

Answer the following questions.

| Question | Answer |
|---|---|
| How many cars has **each** girl picked so far? | |
| How many cars were selected so far **overall**? | |
| What is the total number of cars that were not assigned to a race yet? | |

## Part 5

Help Jennifer and Britney remember which cars should race together. Draw lines between the cars on the opposite page that are paired in the same race. Write the race numbers in the bubbles assigned and shade each pair of cars using a different color.

Color the box with the correct answers for each question below.

Did each girl get the same number of cars?

| Yes | No |
|---|---|

How many cars were leftover unpaired?

| 0 | 2 | 1 |
|---|---|---|

How many cars does each girl has to race with?

| 9 | 8 | 7 |
|---|---|---|

To split evenly, each girl had to receive

| the same number of cars | a different number of cars | doesn't matter how many cars |
|---|---|---|

# Conclusion: The Number of Cars Was Even!

**Even** numbers can be split equally between **two sets**, just like you've seen the cars in this example. Numbers that are not even cannot be split justly, or evenly, in two. The set on the opposite page shows an **even** split, meaning that no car is left unpaired. Can you think of a number of cars that cannot be evenly split in two? Can 3 cars be split evenly in two?

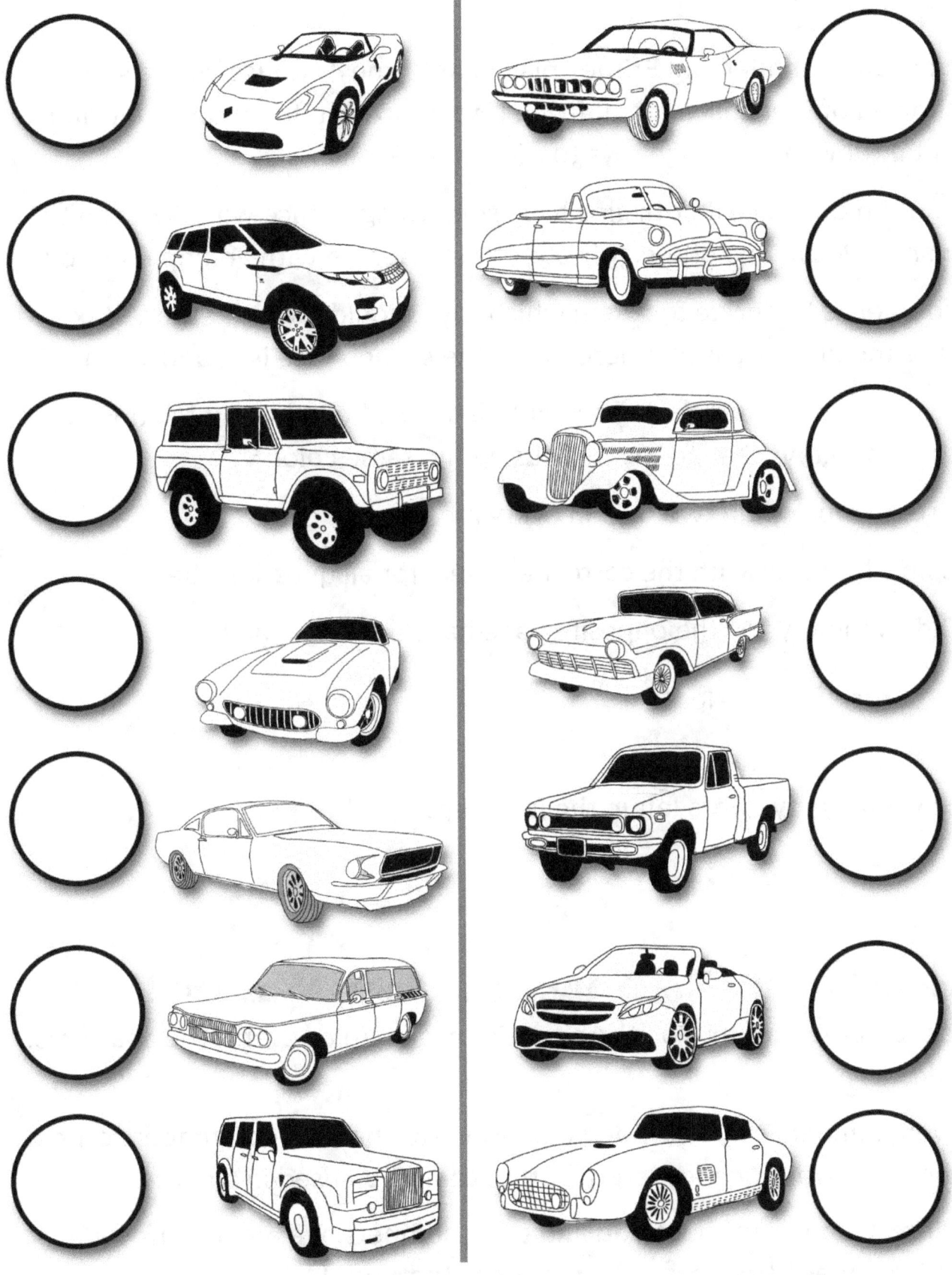

# Worksheet 8

Next day, Jennifer and Britney played at Britney's house where she showed off her collection of flying toys. Help Jennifer and Britney find out if they have enough toys to race them evenly.

You can only race if both girls have something to race with. We don't want to leave one of the girls out of the last race with an unpaired toy.

Your job is to decide for them which toys should race together. Make sure the first race is numbered as 1, the second race is 2, and so on.

Write the race number in each bubble to see how they match up and shade the toys that go together using the same color.

Use **no more than two toys** in every race.

Color the boxes with the correct answers for all questions below.

Did all the toys on the opposite page have a racing match?

| Yes | No |
|---|---|

How many toys were left in the last step?

| 3 | 2 | 1 |
|---|---|---|

# Conclusion: The Number of Toys Was Odd!

Is the number of toys on the opposite page odd or even? The answer is: odd.

Even numbers can be divided equally in two parts. Odd numbers cannot, there is always one toy left over that cannot be used in a pair.

When wondering about **odd** and **even** numbers, we are looking for a split **between two sets only, not more than two!**

# Worksheet 9

Let's find out how many "**2 in 4**" cats we have on the opposite page.

First, circle all the groups of 4 you can find.

In each group, color two of the cats. These cats then belong to the "**2 in 4**" idea.

Write the total number of cats you have colored overall in the blank space below:

_____ cats.

Write the number of cats left uncolored in each group in the blank space below:


_____ cats.

Fill in the following blanks.


In each group there are a total of _____ cats.

_____ of them are colored, and _____ of them were left uncolored.


Shade in the correct answer below.

The number of "**2 in 4**" cats (colored) and the number of left-over cats (not colored) is:

| the same | different |

# Worksheet 10

The toys on the opposite page are forming patterns.

Patterns use sets of items that appear repeatedly.

A question mark was added for a missing toy in every row.

Color the box that shows the missing toy below each pattern.

Find a group of toys that repeats in each row. Circle it and color the toys in it.

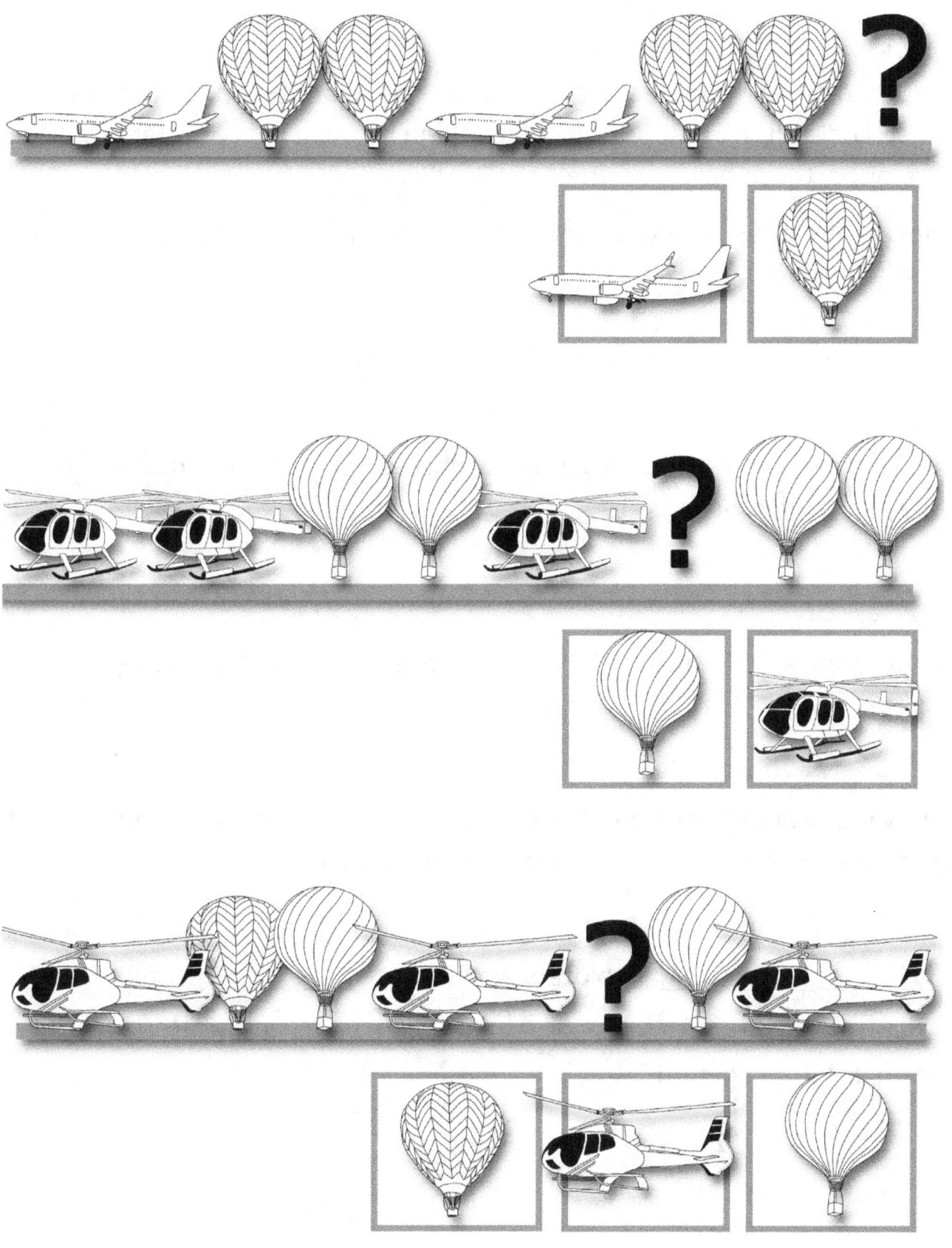

# Worksheet 11

Let's look for how many "**2 in 3**" birds we have on the opposite page. First, circle all the groups of 3 you can find. In each group, color only 2 of the birds, these belong to the "**2 in 3**" idea.

How many birds have you circled in each group? Color the square below containing your answer.

| 1 | 2 | 3 |

Write down the number of **groups of threes** you have found. Use the space provided below.

_____.

How many animals overall have you colored as "**2 in 3**" on the opposite page?

_____.

How many animals were left over uncolored in each group **that were not part of** the "**2 in 3**"? How many are there overall?

_____ uncolored, and _____ overall.

Color the correct answer below. What would you call the left-over animals left uncolored?

| "3 in 4" | "2 in 3" | "1 in 3" |

Color Your Way to Math, Volume 3

# Worksheet 12

Color all the boxes that have

    at least 3 items of any kind,

OR

    at least 1 item that can fly.

# Worksheet 13

Translate the addition from the top row into numbers. Write the numbers in the empty boxes below. Remember from Volume 2 that addition is just like counting everything together as part of one set.

Color the box of cats that has the **largest** counts.

Translate the addition into numbers. Write the numbers in the empty boxes below.

Color the box of cats that has the **largest** counts.

Color Your Way to Math, Volume 3

Translate the addition from the top row into numbers. Write the numbers in the empty boxes below.

Color the box of cats that has the **smallest** counts.

Translate the addition from the top row into numbers. Write the numbers in the empty boxes below.

Color the box of cats that has the **smallest** counts.

# Worksheet 14

Use numbers to fill in the blanks in the text that follows.

## Part 1

There are two toy bins shown below. One on the left and the second on the right. The bin on the left is full. The two bins can hold exactly the same number of toys, one in each slot.

Color the **first** slot of each bin with the same color.

Color the **second** slot of each bin using another color.

Use a third color for the **third** slot of each bin.

We can add _____ more toys to fill the bin on the right.

## Part 2

The two bins below can hold the same number of toys, one in each slot.

Color the **first** slot of each bin with the same color. Use a **second** color to shade the second slot of each bin, and so on.

We can add _____ more toys to fill the bin on the right.

## Part 3

The two bins below can hold the same number of toys, one in each slot.

Color the **first** slot of each bin with the same color. Use a **second** color to shade the second slot of each bin, and so on.

We can add _____ more toys to fill the bin on the right.

## Part 4

This time neither of the bins are full, but each can hold the same number of toys, one in each slot.

Color the **first** slot of each bin with the same color. Use a **second** color to shade the second slot of each bin, and so on.

We can add _____ more toys to fill the bin on the **left**, and we can add _____ more toys to fill the bin on the **right**.

# Worksheet 15

A cube net is given on the opposite page. There are two more on the pages that follow, but for this worksheet we'll only focus on this first net.

Cut out the cube on the opposite page.

Don't assemble your cube just yet. Look at the four cubes listed below. Check a mark next to the cube that is the same as your cutout. Next, color the cubes below that are different than your cutout.

Assemble the cutout from the opposite page into a cube.

If you are not sure how to, go back to the *"Making a Paper Dice"* chapter in Volume 1 for instructions.

Cut out and assemble the two cubes on the pages that follow. Experiment and compare them to the images above. Try to visualize how the faces of every cube connect to each other and see why some match and some don't.

Color Your Way to Math, Volume 3

41

(Cube 1: back side)

(Cube 2: back side)

Color Your Way to Math, Volume 3

(Cube 3: back side)

Draw lines to match each cube with its net.

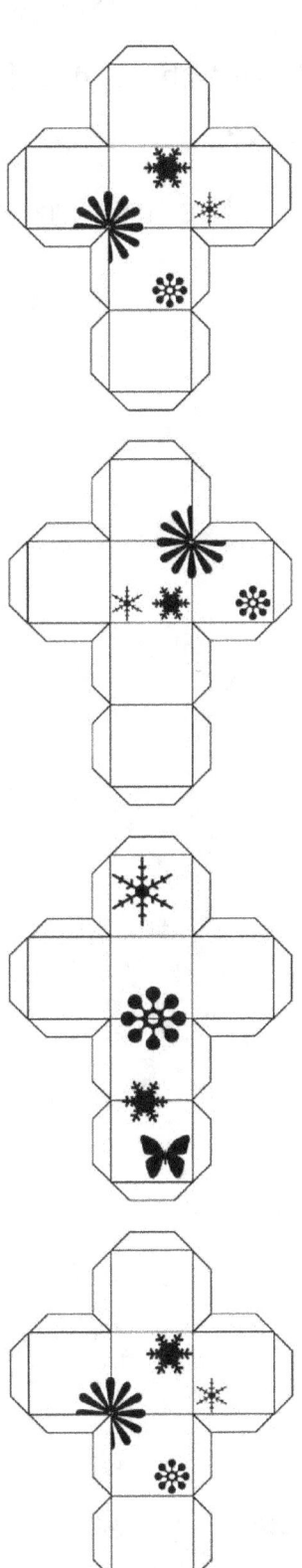

# Worksheet 16

Count the butterflies to answer the addition questions on the opposite page.

Use the first line as an example.

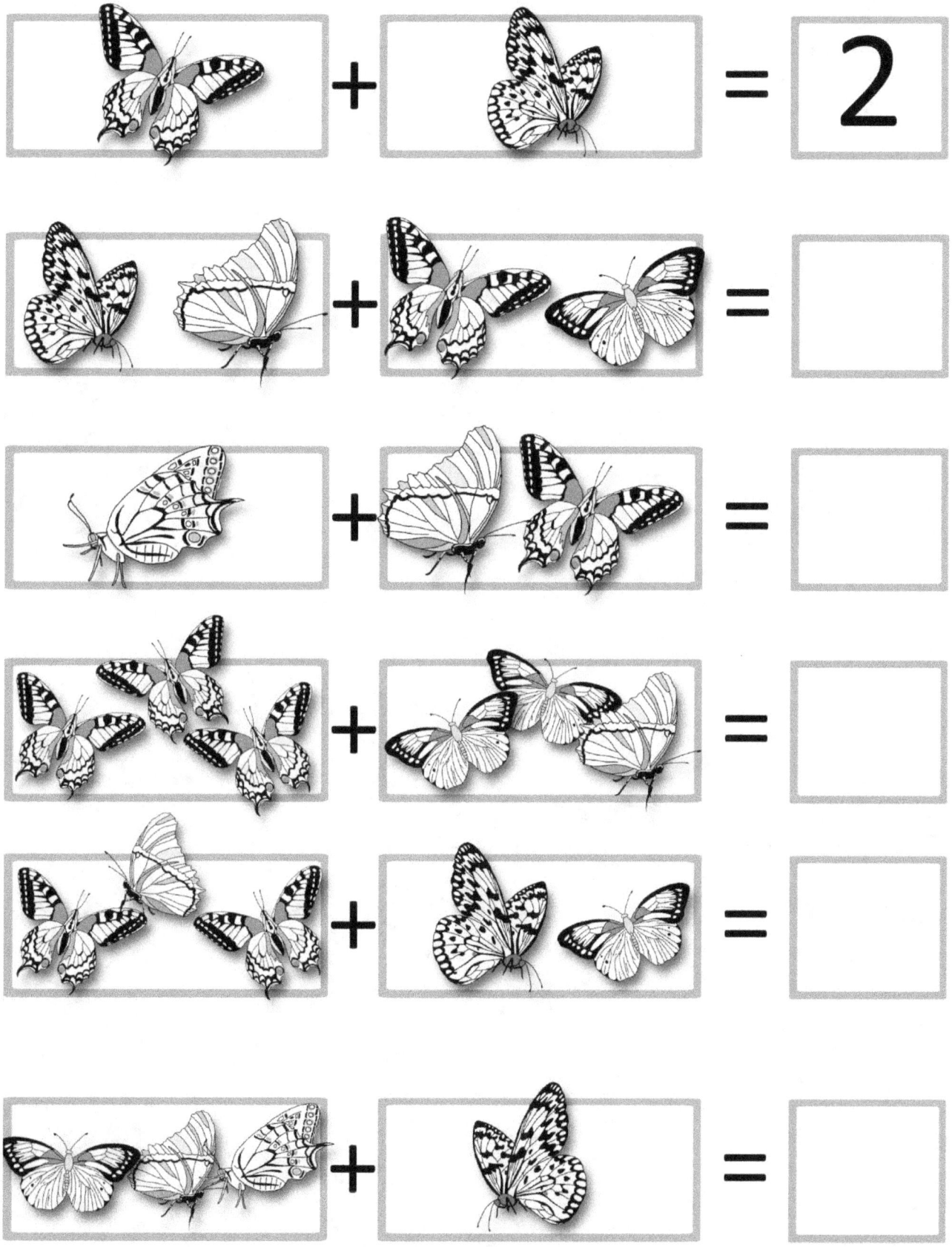

Answers

## Worksheet 1

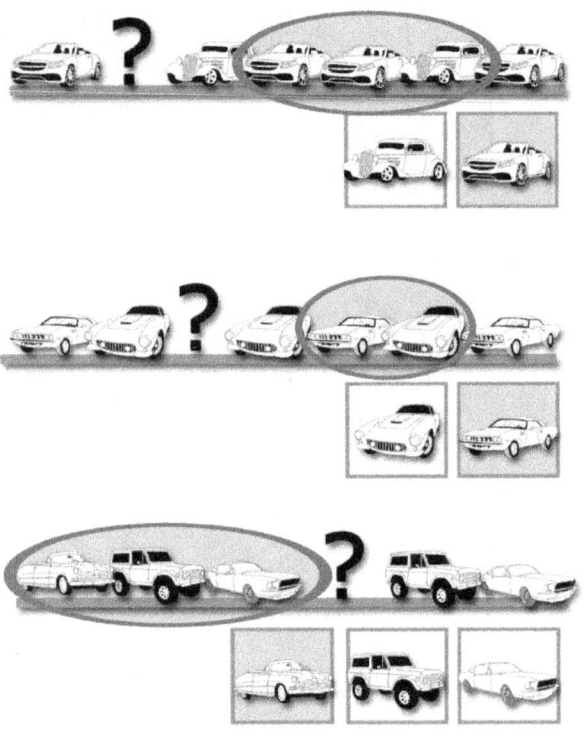

## Worksheet 2

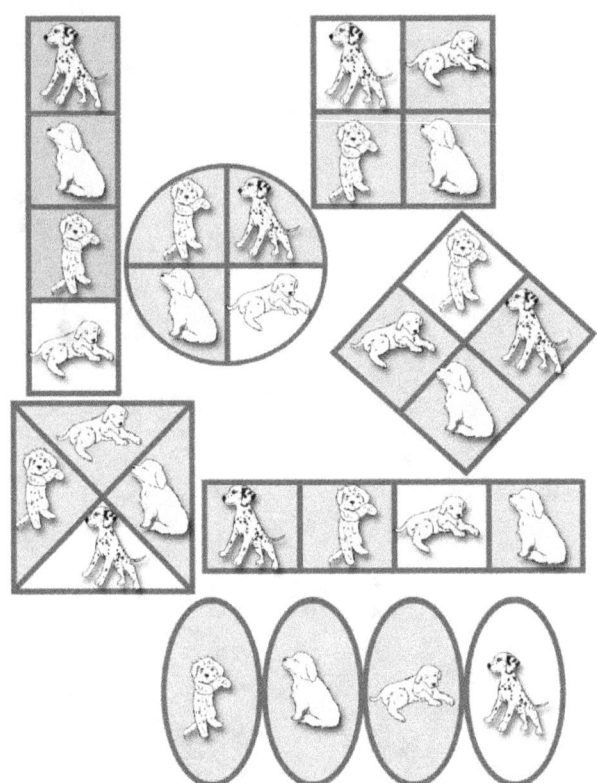

## Worksheet 3

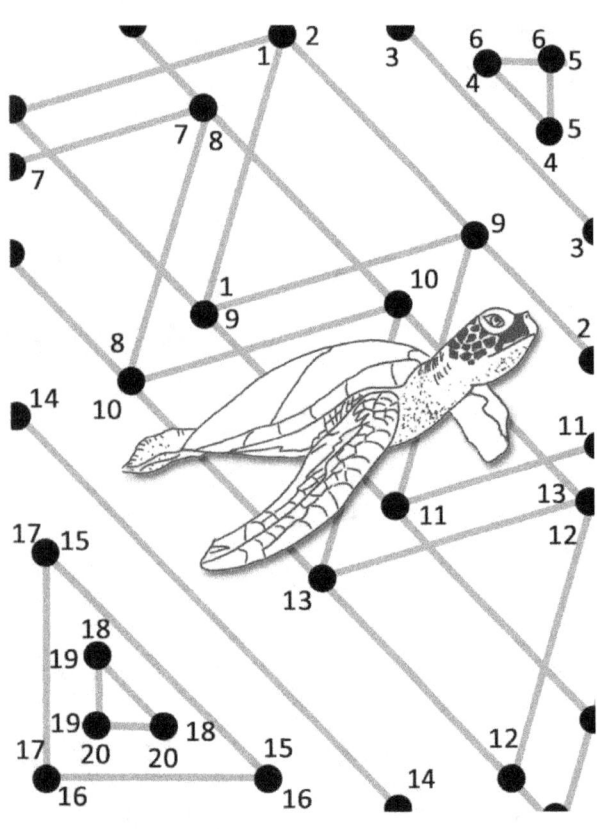

## Worksheet 3, colored

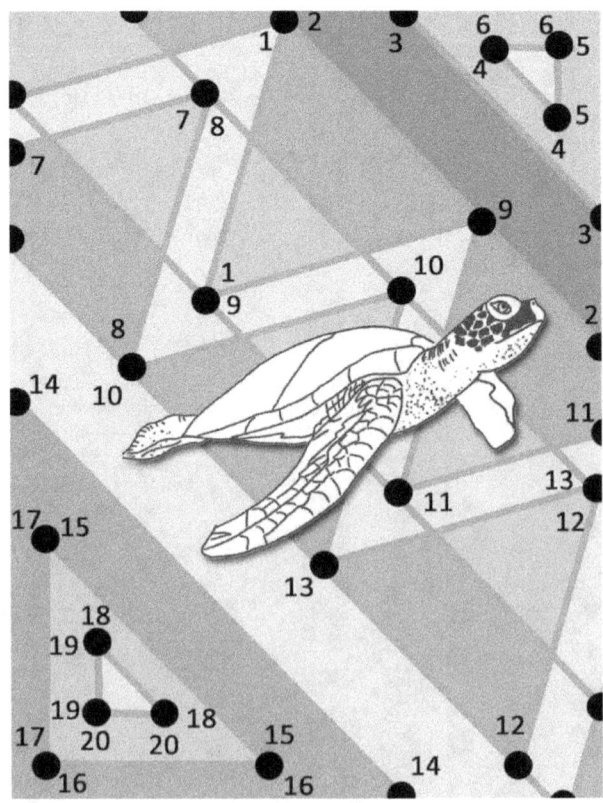

# Worksheet 4

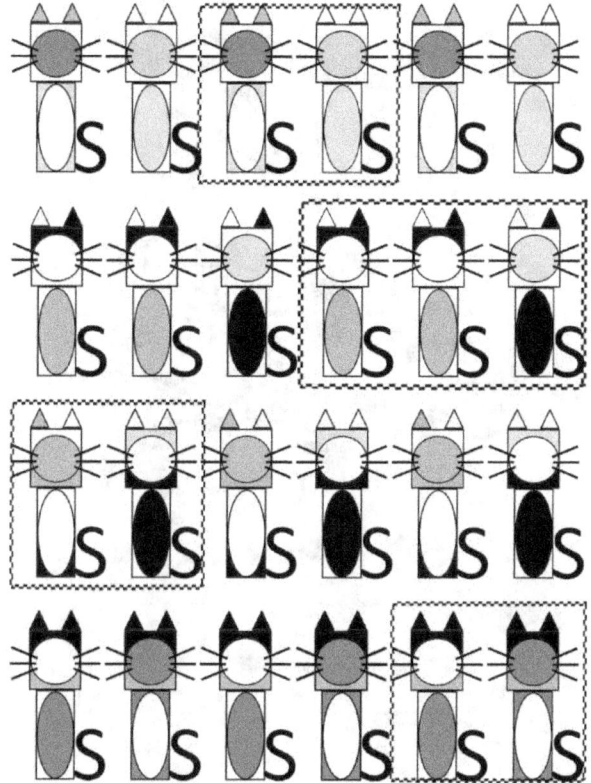

## Worksheet 5

## Worksheet 6

## Worksheet 7, Part 1

## Worksheet 7, Part 2

| Question | Answer |
|---|---|
| How many cars has **each** girl picked so far? | 3 |
| How many cars were selected so far **overall**? | 6 |
| What is the total number of cars that were not assigned to a race yet? | 8 |

## Worksheet 7, Part 3

| Question | Answer |
|---|---|
| How many cars has **each** girl picked so far? | 6 |
| How many cars were selected so far **overall**? | 12 |
| What is the total number of cars that were not assigned to a race yet? | 2 |

## Worksheet 7, Part 4

| Question | Answer |
|---|---|
| How many cars has **each** girl picked so far? | 7 |
| How many cars were selected so far **overall**? | 14 |
| What is the total number of cars that were not assigned to a race yet? | 0 |

## Worksheet 7, Part 5

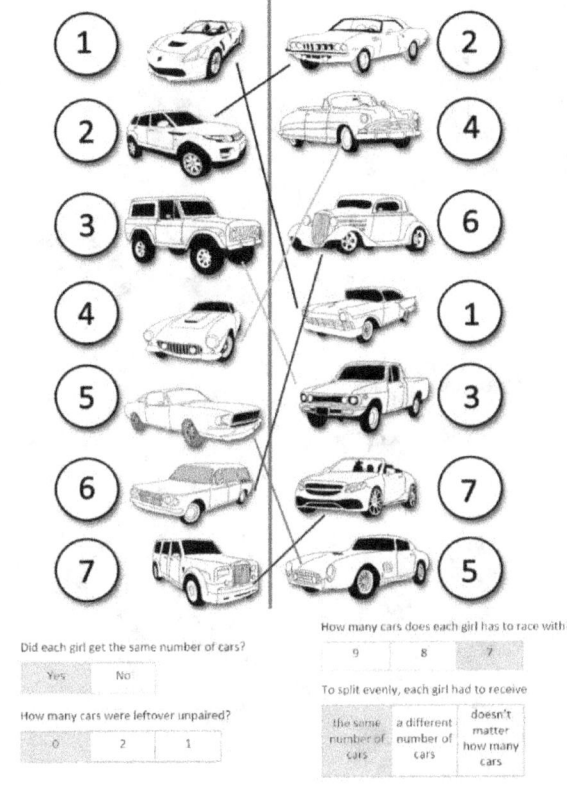

Did each girl get the same number of cars?

Yes    No

How many cars were leftover unpaired?

0    2    1

How many cars does each girl has to race with?

9    8    7

To split evenly, each girl had to receive

the same number of cars | a different number of cars | doesn't matter how many cars

## Worksheet 8

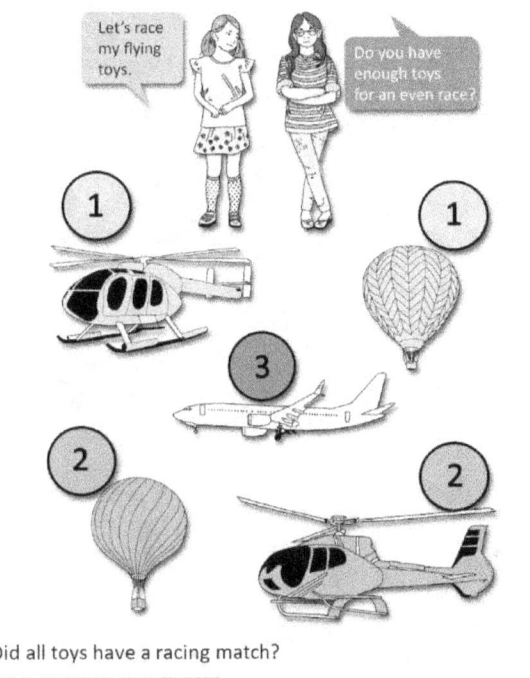

Did all toys have a racing match?

Yes    No

How many toys were left in the last step?

3    2    1

## Worksheet 9

Write down the total number of cats you have colored overall:

6

Write below the number of cats left uncolored in each group:

6

Fill in the blanks.

In each group there are a total of 4 cats. 2 of them are colored, and 2 of them were left uncolored.

Shade in the correct answer below.
The number of "2 in 4" cats (colored) and the number of left-over cats (not colored) is:

the same    different

## Worksheet 10

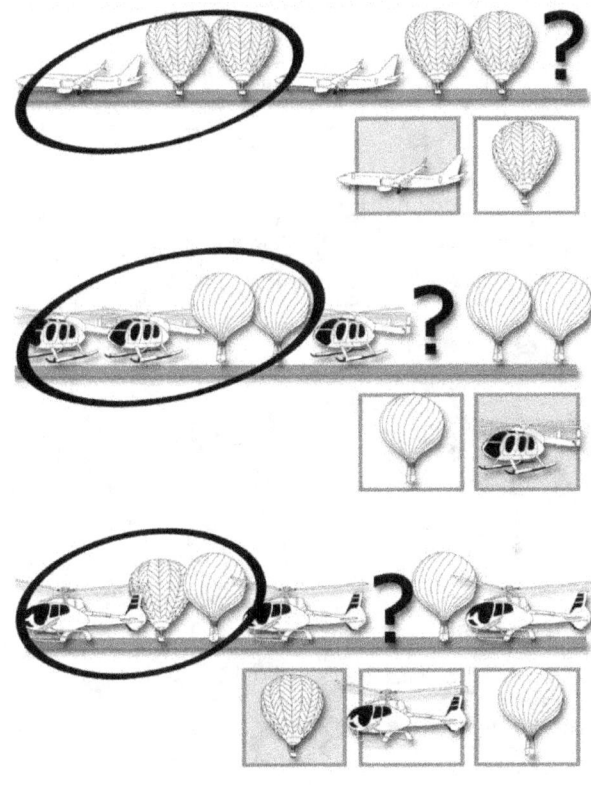

## Worksheet 11

How many animals have you circled in each group? Color the square below containing your answer.

| 1 | 2 | **3** |

Write down the number of **groups of threes** you have found. Use the space provided below.

___3___

How many animals overall have you colored as "**2 in 3**" on the opposite page?

___6___

How many animals were left over uncolored in each group **that were not part of the "2 in 3"**? How many are there overall?

___1___ uncolored, and ___3___ overall.

Color the correct answer below. What would you call the left-over animals left uncolored, that were not part of the "**2 in 3**" sets?

| "3 in 4" | "2 in 3" | **"1 in 3"** |

## Worksheet 12

## Worksheet 13

## Worksheet 14

### Part 1

We can add ___1___ more toys to fill the bin on the right.

### Part 2

We can add ___2___ more toys to fill the bin on the right.

### Part 3

We can add ___3___ more toys to fill the bin on the right.

### Part 4

We can add ___1___ more toys to fill the bin on the left, and we can add ___2___ more toys to fill the bin on the right.

## Worksheet 15

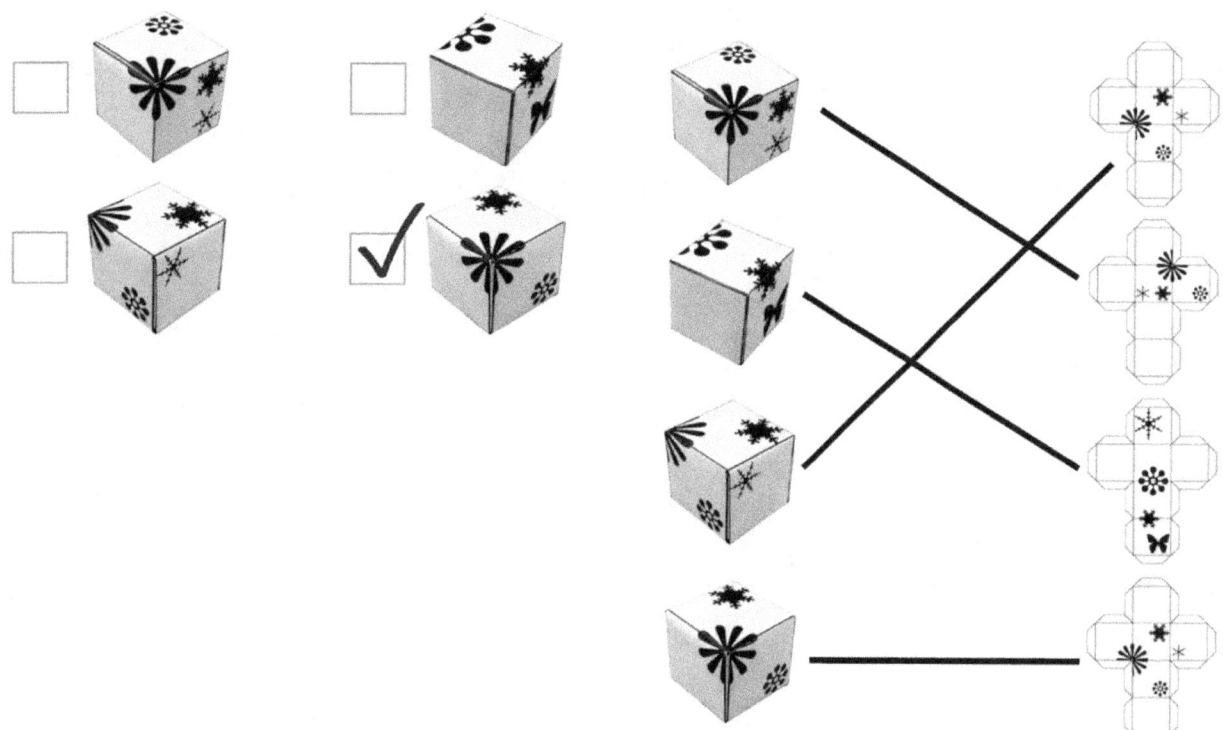

## Worksheet 16

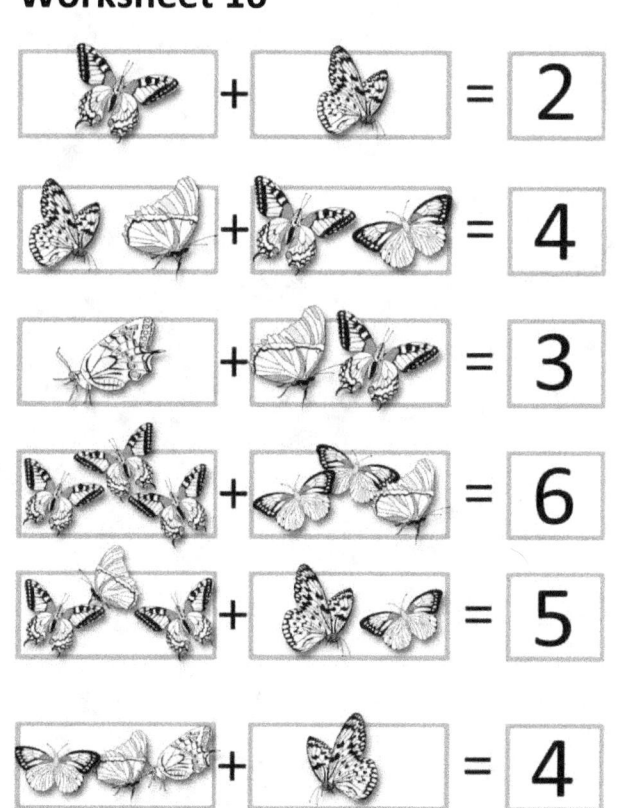

# Errata and Feedback

What are **Errata**?

Sometimes when authors write their books, even after thorough checks, errors can still make their way into a book.

When an error is discovered after printing, it's good to have a place - for example online - where these errors can be recorded so the reader can be warned.

This is especially important for textbooks and manuals.

The place where such errors can be reported is called **Errata** and means a list of published errors. You should always check all your textbooks, including this one, for erratas before you begin your learning.

This book's Errata can be found at:

https://sharpseries.ca/cyw/errata3.html

Comments and suggestions for future editions are always welcome.

# Other Books in This Series

Find out more about **Color Your Way to Math** Volumes at:
https://sharpseries.ca/cyw/cyw3m.html.

Find out more about **Coding with Scratch 3.0**, Workbooks: https://sharpseries.ca/scratch/w.html.

Find out more about **Early Math Concepts** at:

https://sharpseries.ca/em/v1.html.

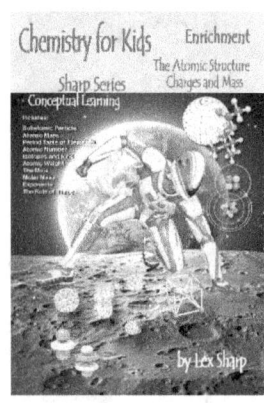

Find out more about **Chemistry for Kids** at:

https://sharpseries.ca/chem/v1.html.

*(All orders are fulfilled by Amazon)*

www.ingramcontent.com/pod-product-compliance
Lightning Source LLC
Chambersburg PA
CBHW081458220526
45466CB00008B/2697